THE COMMON CORE

Clarifying Expectations for Teachers & Students

MATH

Grade K

Created and Presented by
Align, Assess, Achieve

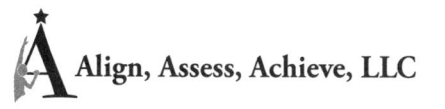

Education

Align, Assess, Achieve, LLC

Align, Assess, Achieve; *The Common Core: Clarifying Expectations for Teachers &*
Students. Grade K

STEM McGraw-Hill is committed to providing instructional materials in Science,
Technology, Engineering, and Mathematics (STEM) that give all students a solid
foundation, one that prepares them for college and careers in the 21st century.

Send all inquiries to:
McGraw-Hill Education
STEM Learning Solutions Center
8787 Orion Place
Columbus, OH 43240

ISBN: 978-002-119894-8
MHID: 0-02-119894-2

Printed in the United States of America.

3 4 5 6 7 8 9 QLM 16 15 14 13 12

Our mission is to provide educational resources
that enable students to become the problem solvers
of the 21st century and inspire them to explore
careers within Science, Technology, Engineering,
and Mathematics (STEM) related fields.

Acknowledgements

This book integrates the Common Core State Standards – a framework for educating students to be competitive at an international level – with well-researched instructional planning strategies for achieving the goals of the CCSS. Our work is rooted in the thinking of brilliant educators, such as Grant Wiggins, Jay McTighe, and Rick Stiggins, and enriched by our work with a great number of inspiring teachers, administrators, and parents. We hope this book provides a meaningful contribution to the ongoing conversation around educating lifelong, passionate learners.

We would like to thank many talented contributors who helped create *The Common Core: Clarifying Expectations for Teachers and Students.* Our team of contributing authors for their intelligence, persistence, and love of teaching; Graphic Designer Thomas Davis, for his creative talents and good nature through many trials; Editors, Sandra Baker, Dr. Teresa Dempsey, and Wesley Yuu, for their educational insight and deep understanding of mathematics; Director of book editing and production Josh Steskal, for his feedback, organization, and unwavering patience; Our spouses, Andrew Bainbridge and Tawnya Holman, who believe in our mission and have, through their unconditional support and love, encouraged us to take risks and grow.

Katy Bainbridge
Bob Holman
Co-Founders
Align, Assess, Achieve, LLC

Executive Editors: *Katy Bainbridge, Bob Holman, Sandra Baker, and Wesley Yuu*
Contributing Authors: *Deborah L. Kaiser, Theresa Mariea, Laura Hance, Ali Fleming, Melissa L. McCreary, Charles L. Brads, Teresa Dempsey, Rebecca Watkins-Heinze, Bob Holman, Wesley Yuu*
Editors: *Jason Bates, Charles L. Brads, Marisa Hilvert, Stephanie Archer*
Graphic Design & Layout: *Thomas Davis; thomasanceldesign.com*
Director of Book Editing & Production: *Josh Steskal*

Introduction

Purpose

The Common Core State Standards (CCSS) provide educators across the nation with a shared vision for student achievement. They also provide a shared challenge: how to interpret the standards and use them in a meaningful way? Clarifying the Common Core was designed to facilitate the transition to the CCSS at the district, building, and classroom level.

Organization

Clarifying the Common Core presents content from two sources: the CCSS and Align, Assess, Achieve. Content from the CCSS is located in the top section of each page and includes the domain, cluster, and grade level standard. The second section of each page contains content created by Align, Assess, Achieve – Enduring Understandings, Essential Questions, Suggested Learning Targets, and Vocabulary. The black bar at the bottom of the page contains the CCSS standard identifier. A sample page can be found in the next section.

Planning for Instruction and Assessment

This book was created to foster meaningful instruction of the CCSS. This requires planning both quality instruction and assessment. Designing and using quality assessments is key to high-quality instruction (Stiggins et al.). Assessment should accurately measure the intended learning and should inform further instruction. This is only possible when teachers and students have a clear vision of the intended learning. When planning instruction it helps to ask two questions, "Where am I taking my students?" and "How will we get there?" The first question refers to the big picture and is addressed with **Enduring Understandings** and **Essential Questions**. The second question points to the instructional process and is addressed by **Learning Targets**.

Where Am I Taking My Students?

When planning, it is useful to think about the larger, lasting instructional concepts as **Enduring Understandings**. Enduring Understandings are rooted in multiple units of instruction throughout the year and are often utilized K-12. These concepts represent the lasting understandings that transcend your content. Enduring Understandings serve as the ultimate goal of a teacher's instructional planning. Although tempting to share with students initially, we do not recommend telling students the Enduring Understanding in advance. Rather, Enduring Understandings are developed through meaningful engagement with an Essential Question.

(continued on next page)

Essential Questions work in concert with Enduring Understandings to ignite student curiosity. These questions help students delve deeper and make connections between the concepts and the content they are learning. Essential Questions are designed with the student in mind and do not have an easy answer; rather, they are used to spark inquiry into the deeper meanings (Wiggins and McTighe). Therefore, we advocate frequent use of Essential Questions with students. It is sometimes helpful to think of the Enduring Understanding as the answer to the Essential Question.

How Will We Get There?

If Enduring Understandings and Essential Questions represent the larger, conceptual ideas, then what guides the learning of specific knowledge, reasoning, and skills? These are achieved by using **Learning Targets**. Learning Targets represent a logical, student friendly progression of teaching and learning. Targets are the scaffolding students climb as they progress towards deeper meaning.

There are four types of learning targets, based on what students are asked to do: knowledge, reasoning/understanding, skill, and product (Stiggins et al.). When selecting Learning Targets, teachers need to ask, "What is the goal of instruction?" After answering this question, select the learning target or targets that align to the instructional goal.

Instructional Goal	Types of Learning Targets	Key Verbs
Recall basic information and facts	Knowledge (K)	Name, identify, describe
Think and develop an understanding	Reasoning/Understanding (R)	Explain, compare and contrast, predict
Apply knowledge and reasoning	Skill (S)	Use, solve, calculate
Synthesize to create original work	Product (P)	Create, write, present

Adapted from Stiggins et al. *Classroom Assessment for Student Learning.* (Portland: ETS, 2006). Print.

Each book contains two types of Enduring Understandings and Essential Questions. The first type, located on the inside cover, relate to the Mathematical Practices, which apply K-12. The second type are based on the domain, cluster, and standard and are located beneath each standard.

Keep in mind that the Enduring Understandings, Essential Questions, and Learning Targets in this book are suggestions. Modify and combine the content as necessary to meet your instructional needs. Quality instruction consists of clear expectations, ongoing assessment, and effective feedback. Taken together, these promote meaningful instruction that facilitates student mastery of the Common Core State Standards.

References

Stiggins, Rick, Jan Chappuis, Judy Arter, and Steve Chappuis. *Classroom Assessment for Student Learning.* 2nd. Portland, OR: ETS, 2006.

Wiggins, Grant, and Jay McTighe. *Understanding by Design, Expanded 2nd Edition.* 2nd. Alexandria, VA: ASCD, 2005.

Page Organization

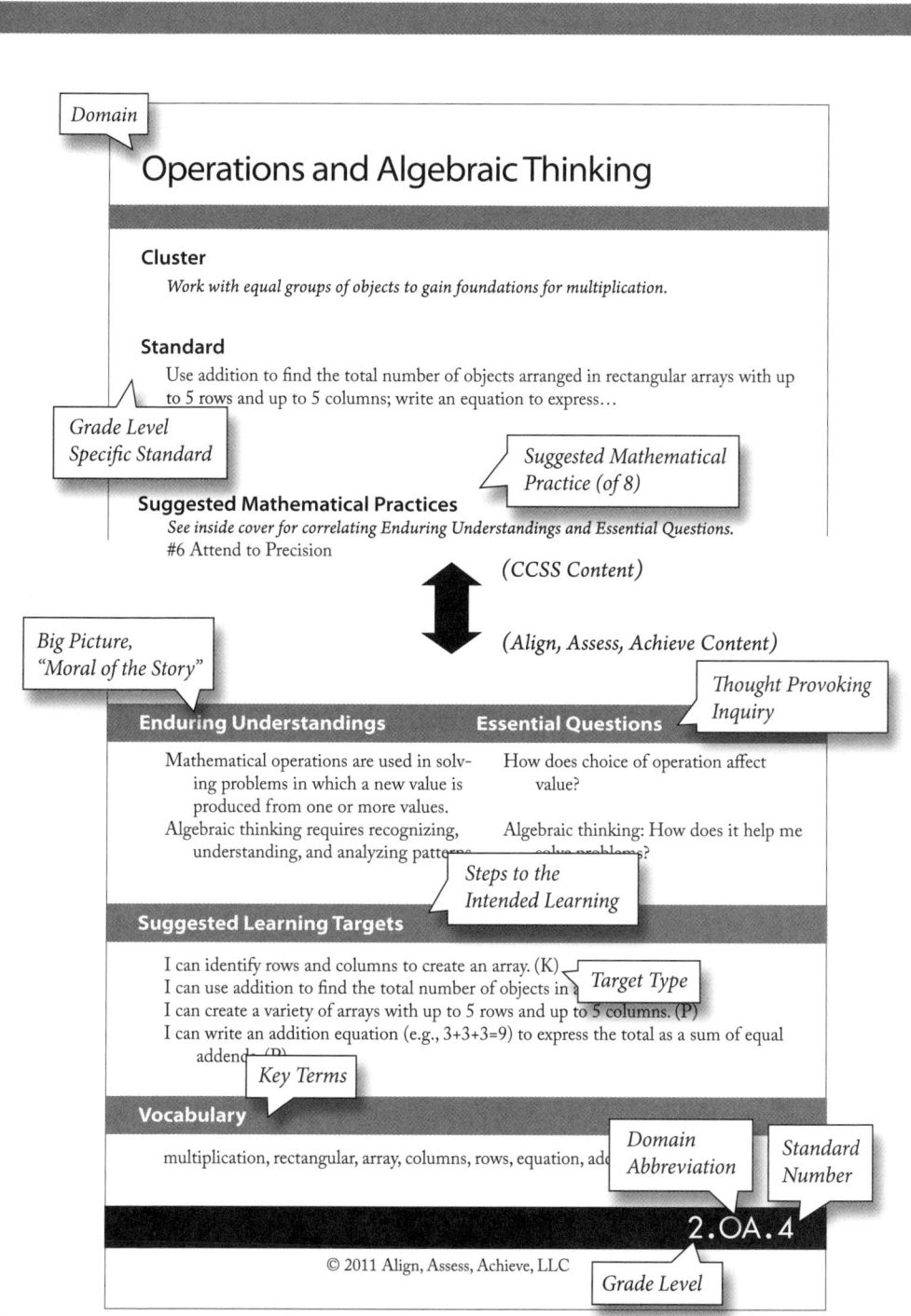

Domain

Operations and Algebraic Thinking

Cluster

Work with equal groups of objects to gain foundations for multiplication.

Standard

Use addition to find the total number of objects arranged in rectangular arrays with up to 5 rows and up to 5 columns; write an equation to express…

Grade Level Specific Standard

Suggested Mathematical Practice (of 8)

Suggested Mathematical Practices

See inside cover for correlating Enduring Understandings and Essential Questions.
#6 Attend to Precision

(CCSS Content)

(Align, Assess, Achieve Content)

Big Picture, "Moral of the Story"

Thought Provoking Inquiry

Enduring Understandings	**Essential Questions**
Mathematical operations are used in solving problems in which a new value is produced from one or more values.	How does choice of operation affect value?
Algebraic thinking requires recognizing, understanding, and analyzing patterns.	Algebraic thinking: How does it help me solve problems?

Steps to the Intended Learning

Suggested Learning Targets

I can identify rows and columns to create an array. (K)
I can use addition to find the total number of objects in a
I can create a variety of arrays with up to 5 rows and up to 5 columns. (P)
I can write an addition equation (e.g., 3+3+3=9) to express the total as a sum of equal addends. (P)

Target Type

Key Terms

Vocabulary

multiplication, rectangular, array, columns, rows, equation, add

Domain Abbreviation

Standard Number

2.OA.4

Grade Level

Mathematical Practices

#1 Making sense of problems and persevere in solving them.

Mathematically proficient students start by explaining to themselves the meaning of a problem and looking for entry points to its solution. They analyze givens, constraints, relationships, and goals. They make conjectures about the form and meaning of the solution and plan a solution pathway rather than simply jumping into a solution attempt. They consider analogous problems, and try special cases and simpler forms of the original problem in order to gain insight into its solution. They monitor and evaluate their progress and change course if necessary. Older students might, depending on the context of the problem, transform algebraic expressions or change the viewing window on their graphing calculator to get the information they need. Mathematically proficient students can explain correspondences between equations, verbal descriptions, tables, and graphs or draw diagrams of important features and relationships, graph data, and search for regularity or trends. Younger students might rely on using concrete objects or pictures to help conceptualize and solve a problem. Mathematically proficient students check their answers to problems using a different method, and they continually ask themselves, "Does this make sense?" They can understand the approaches of others to solving complex problems and identify correspondences between different approaches.

#2 Reason abstractly and quantitatively.

Mathematically proficient students make sense of quantities and their relationships in problem situations. They bring two complementary abilities to bear on problems involving quantitative relationships: the ability to *decontextualize*—to abstract a given situation and represent it symbolically and manipulate the representing symbols as if they have a life of their own, without necessarily attending to their referents—and the ability to *contextualize*, to pause as needed during the manipulation process in order to probe into the referents for the symbols involved. Quantitative reasoning entails habits of creating a coherent representation of the problem at hand; considering the units involved; attending to the meaning of quantities, not just how to compute them; and knowing and flexibly using different properties of operations and objects.

#3. Construct viable arguments and critique the reasoning of others.

Mathematically proficient students understand and use stated assumptions, definitions, and previously established results in constructing arguments. They make conjectures and build a logical progression of statements to explore the truth of their conjectures. They are able to analyze situations by breaking them into cases, and can recognize and use counterexamples. They justify their conclusions, communicate them to others, and respond to the arguments of others. They reason inductively about data, making plausible arguments that take into account the context from which the data arose. Mathematically proficient students are also

(continued on next page)

able to compare the effectiveness of two plausible arguments, distinguish correct logic or reasoning from that which is flawed, and—if there is a flaw in an argument—explain what it is. Elementary students can construct arguments using concrete referents such as objects, drawings, diagrams, and actions. Such arguments can make sense and be correct, even though they are not generalized or made formal until later grades. Later, students learn to determine domains to which an argument applies. Students at all grades can listen or read the arguments of others, decide whether they make sense, and ask useful questions to clarify or improve the arguments.

#4 Model with mathematics.

Mathematically proficient students can apply the mathematics they know to solve problems arising in everyday life, society, and the workplace. In early grades, this might be as simple as writing an addition equation to describe a situation. In middle grades, a student might apply proportional reasoning to plan a school event or analyze a problem in the community. By high school, a student might use geometry to solve a design problem or use a function to describe how one quantity of interest depends on another. Mathematically proficient students who can apply what they know are comfortable making assumptions and approximations to simplify a complicated situation, realizing that these may need revision later. They are able to identify important quantities in a practical situation and map their relationships using such tools as diagrams, two-way tables, graphs, flowcharts and formulas. They can analyze those relationships mathematically to draw conclusions. They routinely interpret their mathematical results in the context of the situation and reflect on whether the results make sense, possibly improving the model if it has not served its purpose.

#5 Use appropriate tools strategically.

Mathematically proficient students consider the available tools when solving a mathematical problem. These tools might include pencil and paper, concrete models, a ruler, a protractor, a calculator, a spreadsheet, a computer algebra system, a statistical package, or dynamic geometry software. Proficient students are sufficiently familiar with tools appropriate for their grade or course to make sound decisions about when each of these tools might be helpful, recognizing both the insight to be gained and their limitations. For example, mathematically proficient high school students analyze graphs of functions and solutions generated using a graphing calculator. They detect possible errors by strategically using estimation and other mathematical knowledge. When making mathematical models, they know that technology can enable them to visualize the results of varying assumptions, explore consequences, and compare predictions with data. Mathematically proficient students at various grade levels are able to identify relevant external mathematical resources, such as digital content located on a website, and use them to pose or solve problems. They are able to use technological tools to explore and deepen their understanding of concepts.

(continued on next page)

#6 Attend to precision.

Mathematically proficient students try to communicate precisely to others. They try to use clear definitions in discussion with others and in their own reasoning. They state the meaning of the symbols they choose, including using the equal sign consistently and appropriately. They are careful about specifying units of measure, and labeling axes to clarify the correspondence with quantities in a problem. They calculate accurately and efficiently, express numerical answers with a degree of precision appropriate for the problem context. In the elementary grades, students give carefully formulated explanations to each other. By the time they reach high school they have learned to examine claims and make explicit use of definitions.

#7 Look for and make use of structure.

Mathematically proficient students look closely to discern a pattern or structure. Young students, for example, might notice that three and seven more is the same amount as seven and three more, or they may sort a collection of shapes according to how many sides the shapes have. Later, students will see 7×8 equals the well remembered $7 \times 5 + 7 \times 3$, in preparation for learning about the distributive property. In the expression $x^2 + 9x + 14$, older students can see the 14 as 2×7 and the 9 as $2 + 7$. They recognize the significance of an existing line in a geometric figure and can use the strategy of drawing an auxiliary line for solving problems. They also can step back for an overview and shift perspective. They can see complicated things, such as some algebraic expressions, as single objects or as being composed of several objects. For example, they can see $5 - 3(x - y)^2$ as 5 minus a positive number times a square and use that to realize that its value cannot be more than 5 for any real numbers x and y.

#8 Look for and express regularity in repeated reasoning.

Mathematically proficient students notice if calculations are repeated, and look both for general methods and for shortcuts. Upper elementary students might notice when dividing 25 by 11 that they are repeating the same calculations over over again, and conclude they have a repeating decimal. By paying attention to the calculation of slope as they repeatedly check whether points are on the line through $(1, 2)$ with slope 3, middle school students might abstract the equation $(y - 2)/(x - 1) = 3$. Noticing the regularity in the way terms cancel when expanding $(x - 1)(x + 1)$, $(x - 1)(x^2 + x + 1)$, and $(x - 1)(x^3 + x^2 + x + 1)$ might lead them to the general formula for the sum of a geometric series. As they work to solve a problem, mathematically proficient students maintain oversight of the process, while attending to the details. They continually evaluate the reasonableness of their intermediate results.

CCSS Grade Level Introduction

In Kindergarten, instructional time should focus on two critical areas: (1) representing and comparing whole numbers, initially with sets of objects; (2) describing shapes and space. More learning time in Kindergarten should be devoted to number than to other topics.

1. Students use numbers, including written numerals, to represent quantities and to solve quantitative problems, such as counting objects in a set; counting out a given number of objects; comparing sets or numerals; and modeling simple joining and separating situations with sets of objects, or eventually with equations such as 5 + 2 = 7 and 7 − 2 = 5. (Kindergarten students should see addition and subtraction equations, and student writing of equations in kindergarten is encouraged, but it is not required.) Students choose, combine, and apply effective strategies for answering quantitative questions, including quickly recognizing the cardinalities of small sets of objects, counting and producing sets of given sizes, counting the number of objects in combined sets, or counting the number of objects that remain in a set after some are taken away.

2. Students describe their physical world using geometric ideas (e.g., shape, orientation, spatial relations) and vocabulary. They identify, name, and describe basic two-dimensional shapes, such as squares, triangles, circles, rectangles, and hexagons, presented in a variety of ways (e.g., with different sizes and orientations), as well as three-dimensional shapes such as cubes, cones, cylinders, and spheres. They use basic shapes and spatial reasoning to model objects in their environment and to construct more complex shapes.

Counting and Cardinality

Cluster

Know number names and the count sequence.

Standard

Count to 100 by ones and by tens.

Suggested Mathematical Practices

See inside cover for correlating Enduring Understandings and Essential Questions.

#7 Look for and make use of structure.

Enduring Understandings	Essential Questions
Counting is a purposeful skill that assigns a number name to an object or set of objects.	What does a number represent? How many are there?

Suggested Learning Targets

I can count to 10 by ones. (K)
I can count to 20 by ones. (K)
I can count to 100 by ones. (K)
I can count to 100 by tens. (K)

Vocabulary

count

K.CC.1

Counting and Cardinality

Cluster

Know number names and the count sequence.

Standard

Count forward beginning from a given number within the known sequence (instead of having to begin at 1).

Suggested Mathematical Practices

See inside cover for correlating Enduring Understandings and Essential Questions.

#7 Look for and make use of structure.

Enduring Understandings	Essential Questions
Counting is a purposeful skill that assigns a number name to an object or set of objects.	What does a number represent? How many are there?

Suggested Learning Targets

I can count to 10. (K)
I can count to 100. (K)
I can count on from a number other than 1 up to 100. (K)

Vocabulary

count

Counting and Cardinality

Cluster

Know number names and the count sequence.

Standard

Write numbers from 0 to 20. Represent a number of objects with a written numeral 0-20 (with 0 representing a count of no objects).

Suggested Mathematical Practices

See inside cover for correlating Enduring Understandings and Essential Questions.

#4 Model with mathematics.

Enduring Understandings	Essential Questions
Counting is a purposeful skill that assigns a number name to an object or set of objects.	What does a number represent? How many are there?

Suggested Learning Targets

I can write numbers 0-10. (K)
I can write numbers 11-20. (K)
I can represent a group of objects with a written numeral 0-20. (S)

Vocabulary

count

Counting and Cardinality

Cluster

Count to tell the number of objects.

Standard

Understand the relationship between numbers and quantities; connect counting to cardinality.

a. When counting objects, say the number names in the standard order, pairing each object with one and only one number name and each number name with one and only one object.

b. Understand that the last number name said tells the number of objects counted. The number of objects is the same regardless of their arrangement or the order in which they were counted.

c. Understand that each successive number name refers to a quantity that is one larger.

Suggested Mathematical Practices

See inside cover for correlating Enduring Understandings and Essential Questions.

#6 Attend to precision.

Enduring Understandings

Counting is a purposeful skill that assigns a number name to an object or set of objects.

Essential Questions

What does a number represent?
How many are there?

Suggested Learning Targets

I can count objects in a group correctly (each object is counted only once) regardless of arrangement and order. (S)

I can say "how many" are in a group after counting all the objects. (S)

If I already know how many are in a group, I can say how many there are (without recounting the whole group) when one more object is added to the group. (S)

I can explain my counting strategy. (R)

Vocabulary

count

Counting and Cardinality

Cluster

Count to tell the number of objects.

Standard

Count to answer "how many?" questions about as many as 20 things arranged in a line, a rectangular array, or a circle, or as many as 10 things in a scattered configuration; given a number from 1–20, count out that many objects.

Suggested Mathematical Practices

See inside cover for correlating Enduring Understandings and Essential Questions.

#6 Attend to precision.

Enduring Understandings	Essential Questions
Counting is a purposeful skill that assigns a number name to an object or set of objects.	What does a number represent? How many are there?

Suggested Learning Targets

I can count objects up to 20 in a variety of arrangements. (S)
I can say "how many" objects are in a group. (S)
I can show the correct number of objects when I am told a number up to 20. (S)

Vocabulary

count

Counting and Cardinality

Cluster

Compare numbers.

Standard

Identify whether the number of objects in one group is greater than, less than, or equal to the number of objects in another group, e.g., by using matching and counting strategies.*

Include groups with up to ten objects.

Suggested Mathematical Practices

See inside cover for correlating Enduring Understandings and Essential Questions.

#2 Reason abstractly and quantitatively.

Enduring Understandings	Essential Questions
Counting is a purposeful skill that assigns a number name to an object or set of objects.	What does a number represent? How many are there?

Suggested Learning Targets

I can say which group has more by matching or counting the number of objects in both groups. (S)

I can say which group has less by matching or counting the number of objects in both groups. (S)

I can say when groups are equal (same as) by matching or counting. (S)

Vocabulary

greater than, less than, equal

Counting and Cardinality

Cluster
Compare numbers.

Standard
Compare two numbers between 1 and 10 presented as written numerals.

Suggested Mathematical Practices
See inside cover for correlating Enduring Understandings and Essential Questions.
#2 Reason abstractly and quantitatively.

Enduring Understandings	Essential Questions
Counting is a purposeful skill that assigns a number name to an object or set of objects.	What does a number represent? How many are there?

Suggested Learning Targets

I can read numerals to 10. (K)
I can tell the values of numbers to 10. (K)
I can compare two numerals between 1 and 10 and say which numeral has a greater value. (R)

Vocabulary

greater than, less than, equal

Operations and Algebraic Thinking

Cluster

Understand addition as putting together and adding to, and understand subtraction as taking apart and taking from.

Standard

Represent addition and subtraction with objects, fingers, mental images, drawings*, sounds (e.g., claps), acting out situations, verbal explanations, expressions, or equations.

Drawings need not show details, but should show the mathematics in the problem. (This applies wherever drawings are mentioned in the Standards.)

Suggested Mathematical Practices

See inside cover for correlating Enduring Understandings and Essential Questions.

#5 Use appropriate tools strategically.

Enduring Understandings	Essential Questions
Mathematical operations are used in solving problems in which a new value is produced from one or more values.	In what ways can operations affect numbers?
Algebraic thinking involves choosing, combining, and applying effective strategies for answering quantitative questions.	How can different strategies be helpful when solving a problem?

Suggested Learning Targets

I can explain addition (putting together and adding to). (R)
I can explain subtraction (taking apart and taking from). (R)
I can identify the mathematical symbols used to show addition and subtraction. (K)
I can show addition and subtraction using objects, fingers, sounds, acting out situations, expressions, and equations. (S)

Vocabulary

addition, add, subtraction, subtract, expression, equation

K.OA.1

Operations and Algebraic Thinking

Cluster

Understand addition as putting together and adding to, and understand subtraction as taking apart and taking from.

Standard

Solve addition and subtraction word problems, and add and subtract within 10, e.g., by using objects or drawings to represent the problem.

Suggested Mathematical Practices

See inside cover for correlating Enduring Understandings and Essential Questions.

#4 Model with mathematics.

Enduring Understandings	Essential Questions
Mathematical operations are used in solving problems in which a new value is produced from one or more values.	In what ways can operations affect numbers?
Algebraic thinking involves choosing, combining, and applying effective strategies for answering quantitative questions.	How can different strategies be helpful when solving a problem?

Suggested Learning Targets

I can add and subtract numbers within 10. (S)
I can solve addition and subtraction word problems using objects and drawings. (S)

Vocabulary

addition, add, subtraction, subtract

K.OA.2

Operations and Algebraic Thinking

Cluster

Understand addition as putting together and adding to, and understand subtraction as taking apart and taking from.

Standard

Decompose numbers less than or equal to 10 into pairs in more than one way, e.g., by using objects or drawings, and record each decomposition by a drawing or equation (e.g., 5 = 2 + 3 and 5 = 4 + 1).

Suggested Mathematical Practices

See inside cover for correlating Enduring Understandings and Essential Questions.

#7 Look for and make use of structure.

Enduring Understandings	Essential Questions
Mathematical operations are used in solving problems in which a new value is produced from one or more values.	In what ways can operations affect numbers?
Algebraic thinking involves choosing, combining, and applying effective strategies for answering quantitative questions.	How can different strategies be helpful when solving a problem?

Suggested Learning Targets

I can decompose (break apart) numbers to 10 using objects or drawings. (R)
I can record the answer using a drawing or equation. (S)

Vocabulary

decompose, equation

K.OA.3

Operations and Algebraic Thinking

Cluster

Understand addition as putting together and adding to, and understand subtraction as taking apart and taking from.

Standard

For any number from 1 to 9, find the number that makes 10 when added to the given number, e.g., by using objects or drawings, and record the answer with a drawing or equation.

Suggested Mathematical Practices

See inside cover for correlating Enduring Understandings and Essential Questions.

#7 Look for and make use of structure.

Enduring Understandings	Essential Questions
Mathematical operations are used in solving problems in which a new value is produced from one or more values.	In what ways can operations affect numbers?
Algebraic thinking involves choosing, combining, and applying effective strategies for answering quantitative questions.	How can different strategies be helpful when solving a problem?

Suggested Learning Targets

I can determine the number to add a given number 1-9 to make 10, and show the answer with a drawing or equation. (S)

Vocabulary

add, make 10, addend, equation

K.OA.4

Operations and Algebraic Thinking

Cluster

Understand addition as putting together and adding to, and understand subtraction as taking apart and taking from.

Standard

Fluently add and subtract within 5.

Suggested Mathematical Practices

See inside cover for correlating Enduring Understandings and Essential Questions.

#8 Look for and express regularity in repeated reasoning.

Enduring Understandings	Essential Questions
Mathematical operations are used in solving problems in which a new value is produced from one or more values.	In what ways can operations affect numbers?
Algebraic thinking involves choosing, combining, and applying effective strategies for answering quantitative questions.	How can different strategies be helpful when solving a problem?

Suggested Learning Targets

I can easily add numbers that add up to 5 or less. (S)
I can easily subtract numbers when the starting number is 5 or less. (S)

Vocabulary

add, subtract

K.OA.5

Number and Operations in Base Ten

Cluster
Work with numbers 11-19 to gain foundations for place value.

Standard
Compose and decompose numbers from 11 to 19 into ten ones and some further ones, e.g., by using objects or drawings, and record each composition or decomposition by a drawing or equation (such as 18 = 10 + 8); understand that these numbers are composed of ten ones and one, two, three, four, five, six, seven, eight, or nine ones.

Suggested Mathematical Practices
See inside cover for correlating Enduring Understandings and Essential Questions.
#4 Model with mathematics.

Enduring Understandings	Essential Questions
Understanding place value can lead to number sense and efficient strategies for computing with numbers.	How does a digit's position affect its value?

Suggested Learning Targets
I can count to 20. (K)
I can use numbers 1-9 to make 10 using objects or drawings (e.g., ten frame, base ten blocks). (S)
I can compose (put together) numbers 11-19 using a ten and some ones, and show my work with a drawing or an equation. (S)
I can decompose (break apart) numbers 11-19 into a ten and some ones, and show my work with a drawing or an equation. (S)

Vocabulary
compose, decompose, equation

K.NBT.1

Measurement and Data

Cluster

Describe and compare measurable attributes.

Standard

Describe measurable attributes of objects, such as length or weight. Describe several measurable attributes of a single object.

Suggested Mathematical Practices

See inside cover for correlating Enduring Understandings and Essential Questions.

#6 Attend to precision.

Enduring Understandings	Essential Questions
Measurement processes are used in everyday life to describe and quantify the world.	Why does "what" we measure influence "how" we measure?
Data displays describe and represent data in alternative ways.	Why display data in different ways?

Suggested Learning Targets

I can describe measurable attributes of objects. (K)
I can describe the measurable attributes of a given object. (S)

Vocabulary

(No applicable vocabulary)

Measurement and Data

Cluster

Describe and compare measurable attributes.

Standard

Directly compare two objects with a measurable attribute in common, to see which object has "more of"/"less of" the attribute, and describe the difference. *For example, directly compare the heights of two children and describe one child as taller/shorter.*

Suggested Mathematical Practices

See inside cover for correlating Enduring Understandings and Essential Questions.

#6 Attend to precision.

Enduring Understandings	Essential Questions
Measurement processes are used in everyday life to describe and quantify the world.	Why does "what" we measure influence "how" we measure?
Data displays describe and represent data in alternative ways.	Why display data in different ways?

Suggested Learning Targets

I can tell which object is longer (or shorter or taller) than the other by comparing them side by side. (S)

I can tell which object can hold more (or less) than the other by filling up one of the objects and pouring it into the other one. (S)

I can tell which object is heavier (or lighter) by lifting one in one hand and the other in my other hand. (S)

I can tell which object is warmer (or colder) than the other by touching them. (S)

Vocabulary

(No applicable vocabulary)

K.MD.2

Measurement and Data

Cluster

Classify objects and count the number of objects in each category.

Standard

Classify objects into given categories; count the numbers of objects in each category and sort the categories by count.*

Limit category counts to be less than or equal to 10.

Suggested Mathematical Practices

See inside cover for correlating Enduring Understandings and Essential Questions.

#6 Attend to precision.

Enduring Understandings	Essential Questions
Measurement processes are used in everyday life to describe and quantify the world.	Why does "what" we measure influence "how" we measure?
Data displays describe and represent data in alternative ways.	Why display data in different ways?

Suggested Learning Targets

I can sort (classify) objects into categories (groups). (R)
I can determine the number of objects in each category. (S)
I can sort the categories by number or count. (R)

Vocabulary

(No applicable vocabulary)

Geometry

Cluster

Identify and describe shapes (squares, circles, triangles, rectangles, hexagons, cubes, cones, cylinders, and spheres).

Standard

Describe objects in the environment using names of shapes, and describe the relative positions of these objects using terms such as *above, below, beside, in front of, behind,* and *next to.*

Suggested Mathematical Practices

See inside cover for correlating Enduring Understandings and Essential Questions.

#6 Attend to precision.

Enduring Understandings	Essential Questions
Geometric attributes (such as shapes, lines, angles, figures, and planes) provide descriptive information about an object's properties and position in space and support visualization and problem solving.	How does geometry better describe objects?

Suggested Learning Targets

I can find and name shapes (e.g., square, circle, triangle) in my environment. (K)

I can describe the position of objects as above, below, beside, in front of, behind, and next to. (R)

Vocabulary

shapes

K.G.1

Geometry

Cluster

Identify and describe shapes (squares, circles, triangles, rectangles, hexagons, cubes, cones, cylinders, and spheres).

Standard

Correctly name shapes regardless of their orientations or overall size.

Suggested Mathematical Practices

See inside cover for correlating Enduring Understandings and Essential Questions.

#8 Look for and express regularity in repeated reasoning.

Enduring Understandings	Essential Questions
Geometric attributes (such as shapes, lines, angles, figures, and planes) provide descriptive information about an object's properties and position in space and support visualization and problem solving.	How does geometry better describe objects?

Suggested Learning Targets

I can name shapes correctly. (K)

I can name shapes correctly even when their size and orientation is unusual or different. (R)

Vocabulary

shapes

K.G.2

Geometry

Cluster

Identify and describe shapes (squares, circles, triangles, rectangles, hexagons, cubes, cones, cylinders, and spheres).

Standard

Identify shapes as two-dimensional (lying in a plane, "flat") or three-dimensional ("solid").

Suggested Mathematical Practices

See inside cover for correlating Enduring Understandings and Essential Questions.

#3 Construct viable arguments and critiques the reasoning of others.

Enduring Understandings	Essential Questions
Geometric attributes (such as shapes, lines, angles, figures, and planes) provide descriptive information about an object's properties and position in space and support visualization and problem solving.	How does geometry better describe objects?

Suggested Learning Targets

I can define two-dimensional as being flat. (K)
I can define three-dimensional as being solid. (K)
I can identify two-dimensional shapes. (K)
I can identify three-dimensional shapes. (K)

Vocabulary

shapes

K.G.3

Geometry

Cluster

Analyze, compare, create, and compose shapes.

Standard

Analyze and compare two- and three-dimensional shapes, in different sizes and orientations, using informal language to describe their similarities, differences, parts (e.g., number of sides and vertices/"corners") and other attributes (e.g., having sides of equal length).

Suggested Mathematical Practices

See inside cover for correlating Enduring Understandings and Essential Questions.

#3 Construct viable arguments and critique the reasoning of others.

Enduring Understandings	Essential Questions
Geometric attributes (such as shapes, lines, angles, figures, and planes) provide descriptive information about an object's properties and position in space and support visualization and problem solving.	How does geometry better describe objects?

Suggested Learning Targets

I can describe a shape by telling things like the number of sides, number of vertices (corners), and other special qualities. (R)

I can compare two-dimensional shapes and describe their similarities and differences. (R)

I can compare three-dimensional shapes and describe their similarities and differences. (R)

Vocabulary

shapes, two-dimensional, three-dimensional

K.G.4

Geometry

Cluster
Analyze, compare, create, and compose shapes.

Standard
Model shapes in the world by building shapes from components (e.g., sticks and clay balls) and drawing shapes.

Suggested Mathematical Practices
See inside cover for correlating Enduring Understandings and Essential Questions.

#4 Model with mathematics.

Enduring Understandings	Essential Questions
Geometric attributes (such as shapes, lines, angles, figures, and planes) provide descriptive information about an object's properties and position in space and support visualization and problem solving.	How does geometry better describe objects?

Suggested Learning Targets

I can build shapes from materials in my environment. (P)
I can draw shapes in my environment. (P)

Vocabulary

shapes

Geometry

Analyze, compare, create, and compose shapes.

Standard

Compose simple shapes to form larger shapes. *For example, "Can you join these two triangles with full sides touching to make a rectangle?"*

Suggested Mathematical Practices

See inside cover for correlating Enduring Understandings and Essential Questions.

#4 Model with mathematics.

Enduring Understandings	Essential Questions
Geometric attributes (such as shapes, lines, angles, figures, and planes) provide descriptive information about an object's properties and position in space and support visualization and problem solving.	How does geometry better describe objects?

Suggested Learning Targets

I can put shapes together to make new shapes. (R)
I can name the new shape that results from composing two simple shapes. (K)

Vocabulary

shapes, compose

K.G.6